YOUR KNOWLEDGE HAS VALUE

- We will publish your bachelor's and
  master's thesis, essays and papers

- Your own eBook and book -
  sold worldwide in all relevant shops

- Earn money with each sale

Upload your text at www.GRIN.com
and publish for free

Bibliographic information published by the German National Library:

The German National Library lists this publication in the National Bibliography; detailed bibliographic data are available on the Internet at http://dnb.dnb.de .

Imprint:

Copyright © 2015 GRIN Verlag, Open Publishing GmbH
Print and binding: Books on Demand GmbH, Norderstedt Germany
ISBN: 978-3-668-03878-3

This book at GRIN:

http://www.grin.com/en/e-book/304096/overview-of-basics-and-types-of-fermentation

Manoj Parakhia, R.S. Tomar, B.A. Golakiya

Overview of Basics and Types of Fermentation

GRIN Publishing

Basic and types of Fermentation

By

Prof. Manoj. V. Parakhia, Rukam S. Tomar and B. A. Golakiya

Index

1. General concept of industrial microbiology

- **Introduction**

> Industrial Microbiology deals with all type of Microbiology which has an economic impact.
> Industrial Microbiology concern with…
> (1) Isolation & Identification of Microbes from natural environment such as soil & water.
> (2) Optimizing the cultural conditions required for obtaining rapid & Massive growth of these organism in laboratory & in frequenter.
> Thus the fermentation industry is a part of industrial Microbiology.

1.1 Concept of Fermentation

> The term "Fermentation" is derived from Latin Verb "Fervere" means to boil.
> Originally fermentation referred to the bubbling, observed when sugar or starchy material caveat into alcoholic beverages.
> Here the bubbling is observed due to the production of Co_2 gas.
> Later on the term "fermentation" was applied to the process in which alcohol was formed from sugar.
> Later on Pasteur described fermentation as those anaerobic process through which microorganism obtained energy for growth in the absence of oxygen.
> But, today Fermentation is broadly use for both, aerobic & anaerobic metabolic activity of microorganism in which specific chemical changes are brought in an organic substrate.
> According to industrial Microbiology, the word fermentation includes almost any process Mediated by or involving Microbes in which a product of economic value produce.
> The biochemical meaning of termination is the generation of energy by the catabolism of organic compounds.

1.2 The range of Fermentation Processes

> This answer include following points.
> (A) Introduction.
> (B) Microbial biomass
> (C) Microbial enzyme.
> (D) Microbial Metabolites
> (E) Recombinant Products
> (F) Transformation Processes

A) Introduction:

- ➤ There are five Major group of commercially important fermentation.
  - Microbial biomass
  - Microbial Enzymes
  - Microbial Metabolites
  - Recombinant Product
  - Transformation Process.

(B) Microbial biomass

- ➤ Commercial production of Microbial biomass may be divided into two major processes.
  - (A) The Production of yeast to be used in baking industry.
  - (B) Production of microbial cells to be sued as food for human & animal (SCP)
- ➤ The production of Banker's yeast is started before 1900s.
- ➤ Yeast was produced as human food in care many during First World War.
- ➤ The Production of microbial biomass as food of animal was established in the 1970. These process based on Hydrocarbon feed stocks.
- ➤ Later on ICI Plo & Rank Hovis McBougal established a process for the production of fungal biomass for human food.

(C) Microbial Enzymes

- ➤ Enzyme commercially produced form plant, animals & Microbial origin.
- ➤ However Microbial enzyme is produced in large quantities by establishment of fermentation techniques.
- ➤ It easier to improve the productivity of microbial system as compared to plants & animals.
- ➤ With the help of R-DNA tech, it is possible produced animal origin enzyme by the Microorganism.

Table 1.1 Summarize the use of microbial enzyme.

Industry	Application	Enzyme	Sources
Laundry	Detergent	Protease, Lipase	Bacteria
Leather	Decaying, baiting	Protease	Bacterial fungi
Dairy	Stabilization of evaporated Milk	Protease	Fungal.
Fruit Juice	Oxygen removal	Glucose oxidase	Fungal
Pharmaceutical	Antiblood Clotting	Streptokinase	Bacteria
Soft Drink	Stabilization	Glucose Oxidase	Fungal
Meat Tenderization	Tenderization	Protease	Fungal

- ➤ Enzyme production is controlled by microbes, therefore in order to improved productivity we have to modify this control.
- ➤ Induction of enzyme synthesis is carried cut by adding inducer in medium.
- ➤ While repression control of enzyme is removed by mutation or recombination. Tech.
- ➤ Also the number of gene that encode for the enzyme can be increased by R-DNA technology.
- ➤ Aspects of strains improvement are also included for improvement of enzyme synthesis.

(D) Microbial Metabolites.

- ➤ The growth of microbial culture is divided into four stages.
    - (A) Lag Phase
    - (B) Log Phase
    - (C) Stationary Phase
    - (D) Decline Phase

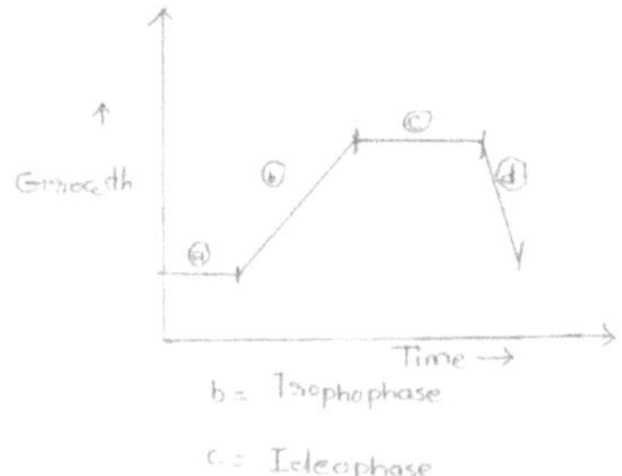

- ➤ The behavior of a culture may also be described according to the product, which they product, which they produced during the various stage of growth.

Fig.1.1 Growth phases of Bacteria

(A) Primary Metabolites

- ➤ The product, which produced during log phase are essential for the cell growth like N. Acid, proteins, lipids, carbohydrates etc.
- ➤ These products are referred as "Primary Metabolites" & the phase of growth in which these products are produced is called "Trophophase".
- ➤ This Trophophase is equivalence to log phase.
- ➤ Many primary metabolites have great economic important & produced by fermentation.

Table 1.2 Metabolites produced by bacteria.

Primary Metabolites	Significance
Ethanol	Ingredient for alcoholic drink
Glutamic acid	Flavour enhancer
Nucleotide	Flavour enhancer
Phenylalanine	Sweetener
Polysaccharide	Use in food industries

➤ Microbiologist Modified the Microbes in such a way that they can produce high amount of primary metabolites.

(B) Secondary Metabolites

➤ During stationary phase some Microbes produce a compound which is not produced during trophophase & which have no function in the cell metabolism.
➤ This compound is reoffered as "Secondary Metabolites" & the phase in which they are synthesized called "idiophase".
➤ Secondary metabolites produced at slow growth rate of culture.
➤ They are generally produced from intermediates & products of primary metabolism.
➤ Secondary metabolites generally produced by filamentous bacteria, fungi & sporulating bacteria.
➤ Many secondary metabolites have antibacterial & antimicrobial activities, other is specific enzymes inhibitors, some are growth promoters & many have Pharmacological properties.
➤ Thus, the product of secondary metabolism formed the basis of number of fermentation industry.

(C) Recombinant Products.

➤ R-DNA tech is now a day used for the production of fermentation products.
➤ The genes of higher organism incorporated in Microbial cell in such a way that they can able to produce the product of this gene.
➤ Following Microbes are asked e-Coli, Saccharomyces cerevisiae, filamentous fungi.
➤ Product produced by R-DNA tech. include – Interferon's, insulin, factor VIII & IX, epidermal growth factor, calf thymosin & bovine stomatostatin.
➤ Following factor is imp for such products.
   (1)   Secretion of product by microbial host.
   (2)   Minimization of degradation of products
   (3)   Control of the onset of synthesis during fermentation.
   (4)   Maximization the operation of the foreign gene.

(D) Transformation Process

➤ Microbial cell may use to covert one compound into structurally related other compound which have more financial value.
➤ Microbes can behave as catalyst & carry out specific change in compound.
➤ Microbial processes are more specific & carry out at normal temperate & pressure then purely chemical method.

➤ Following reduction carried out by microbes.

  - Dehydration   - decarburization
  - Oxidation   - Amination
  - Hydroxylation   - Deamination
  - Dehydrogenation   - isomerisation.

➤ The most studies Microbial transformation is production of vinegar from ethanol.

➤ They also used to produce antibiotics, prostaglandins, steroids.

➤ But only disadvantage of this process is that, they required high amount of biasness.

➤ To solve this problem, immobilized cell & immobilized enzyme. Can be used, which carry reaction on inert support & used for many times.

1.3 The Component part of a Fermentation process.

➤ The established fermentation process may be divided into six basic component parts :

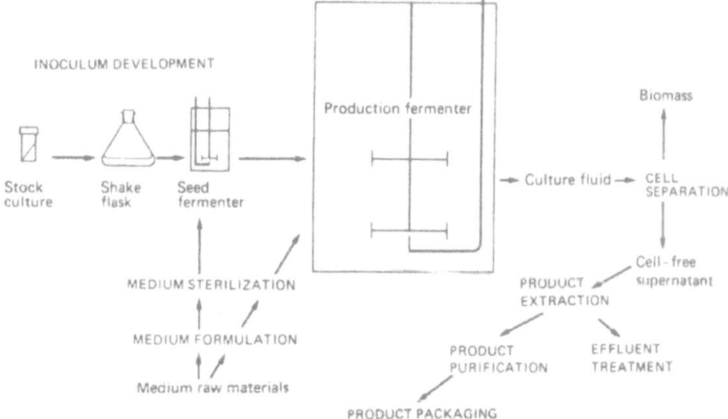

FIg.1.2 Component part of a Fermentation

(1) The Formulation of medium:-

    It is important for inoculums development & the production fermenter.

(2) The sterilization of medium, fermenter & other necessary equipment.

(3) The production of active, pure culture for inoculums.

(4) The growth of organism in fermenter under optimum condition for product Fermenter.

(5) The extraction & purification of product.

(6) Disposal of effluents produced by the process.

➢ Above figure illustrate the inter relationship between the six components part of fermentation process.
➢ But research & development can gradually improved the overall efficiency of the fermentation.

Some time according to the product component will change.

2. The Medium for the industrial fermentation

➢ This answer should include following points.
  (A) Introduction
  (B) Types of Media
  (C) Ideal characteristics of media
  (D) Component of media
      ➢ Water
      ➢ Energy source
      ➢ Minerals
      ➢ Growth factors
      ➢ Buffer
      ➢ Precursor & Metabolic regulators
      ➢ Antifoam agent.

(A) Introduction

➢ Choice of good medium is very important for the success of industrial fermentation.
➢ The Medium supplied nutrient for the growth, energy & building block of the cell substances and biosynthesis of fermented product.

(B) Type of Media

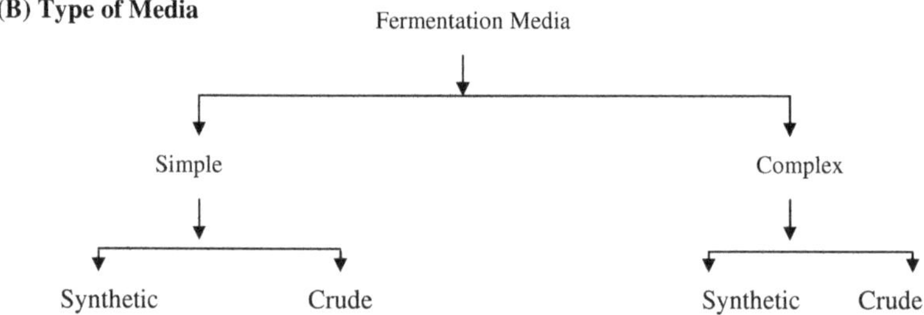

➢ Some autotrophic bacterial required very simple medium which contain water, inorganic salt & Nitrogen source.
➢ While highly fastidious organism (e.g. lactic acid bacteria) require complex medium.
➢ **Synthetic medium: -** Constituent of media is specifically defined & known compound.

- **Advantage :-**
➢ Radio labeling of substrate ⟶ study of metabolic pathway lead to product formation.
➢ Forming is not observed.
➢ Product recovery & purification is very easy.

- **Disadvantage :**
➢ It is very expensive .
➢ **Crude Media :-** The source of nutritional compound is not clefined, but provide excess nutrient & growth factor then defined medium.
➢ Mostly crude media is Agriculture by product eg. Soyabean.

- **Advantage :-**
➢ It is very cheaper.

(C) Ideal Characteristics of Media

➢ Buffering capacity
➢ Lack of foam product
➢ Control of oxidation & reduction potential.
➢ Inhibite the growth of contaminant.
➢ Should maintain genetic stability of organism
➢ Easily sterilizable
➢ Easy to recovery product
➢ Allow the growth of organism in proper morphological structure.
➢ Economicaly feasible.

2.1 Component of Media
(1) Water

➢ Water is a major component of fermentation media.
➢ Provide minerals, metals which act as activator or inhibitor of microbial growth.
➢ Also affect the flavour of product ex; alcohol product.
➢ It also used for cleaning, cooling, steam generation as well as product extraction.
➢ Industries must have good availability of water.

(2) Energy Source

➢ The energy required for growth come from oxidation of medium component or from light.
➢ The common energy source is the source of carbon like carbohydrates, lipid protein & nitrogen.

(A) Carbon Source
▪ **Factor affect the choice of Carbon source.**
   (1) The rate of metabolism: - It affects the formation of biomass, primary & secondary metabolites.
   (2) When product formed directly from carbon source, then cost of "C" source decide the price of product.
   (3) Purity of carbon source. Affect the choice of substrate.
   (4) Method of sterilization eg. Starch becomes viscous.
   (5) The choice of substrate may also affected by government legislation.

▪ **Example of common used carbon source.**

   (1) Carbohydrates

   (a) **Starch:** Obtained from maize grain.
      ➢ Starch is hydrolysed by diluted acid & enzymes to give variety of glucose preparation.

   (b) **Sucrose:** Obtained from sugar cane & sugar beet.
      ➢ Commenly used in media as impure form as beet or cane molasses.

   (c) **Molasses:** It is concentrated syrup
      ➢ It is concentrated syrup recovered from the sugar refinning process

Black Strap Molasses

      ➢ Prepared from sugarcane juice after repeated crystalization of sucrose.
      ➢ Contain total 52% sugar : 30% Sucrose + 22% invert sugar.

High Test Molasses

> ➢ For such molasses, sugar juice is partially hydrolysed to monosaccharride by heat & acid, then nutralized it without removal of sugar.
> ➢ Produced during overproduction of sugar cane.
> ➢ It cantain 70 to 75% sugar.

Beet Molasses

> ➢ Produced similar to cane molasses, it dificent in biotin.

Corn Molasses

> ➢ Molasses resulting from the manufacture of dextrose from corn starch.
> ➢ Contain high amount of salt & 60% sugar.

(d) **Oils & Fats :**
> ➢ Vegetable oil may be used as carbon source.
> ➢ They contain 2.4 times more energy then sugar, required less amount.
> ➢ Also have antifoam properaties.

(e) **Hydrocarbon & their derivatives :**
> ➢ n-Alkane are used for production of organic acid & organic acid, vitabmisn etc.
> ➢ methane & methanol uscd for biomass production
> ➢ By product of petrolium industries.
> ➢ They have twice time carbon & three times energy them same amount of sugar.
> ➢ Use for production of single cell protein.

(B) Nitrogen Source :

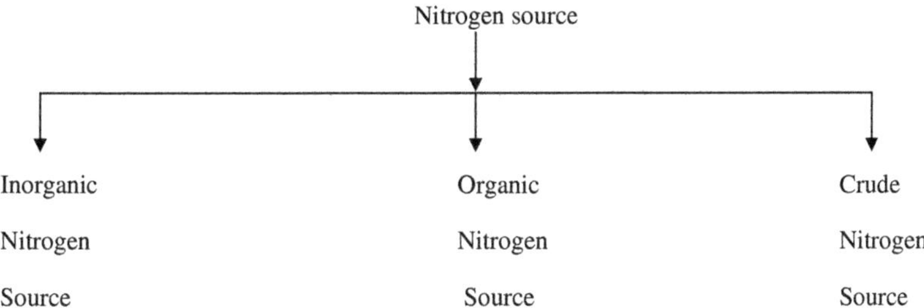

Inorganic Nitrogen Source :-

Ammonia gas, Ammonium salt, nitrates etc.

 ➢ Utilization of ammonium salt generates acid, while nitrates create alkaline condition.

Organic Nitrogen Source :-

Amino acid, protein, urea.

Crude nitrogen source :-

Corn steep liquor, soybean, peanut meal, cotton seed meal.

Factor affect the choice of nitrogen source :-

 ➢ Choice is depend on type of organism & product.
Ex. : Production of Antibiotics can be inhibited by rapid utilization of nitrogen source ex. NO_3, NH_4 Amino Acid.

 ➢ Some N Source also cause problem in influence tratement.

(3) Minerals :

 ➢ All microbes required minerals for there growth & metabolism.
 ➢ Mg. P, K. Ca, are added additionaly because they are essential.
 ➢ Other Cu, Co, Fe, Mn are les essential, present in Medium as impurity of raw material.
 ➢ Act as Co-Factor of enzymes.
 ➢ Sometimes chelating agent like EDTA is also added.

(4) Growth factor:-

 ➢ Some microbes can not synthesized certain cell component & added in the medium called growth factor"
 ➢ Growth factor are generally vitamins, specific amino acid, fatty acid, sterols.
 ➢ Crude media contain such growth factor, so no need to add.

(5) Buffer :-

P^H of the medium affects the productivity of process.

- ➢ Buffer is added to control P^H
- ➢ $CaCO_3$ & phosphate buffer mantained the P^H
- ➢ Buttering capacity also provided by protein & amino acid.
- ➢ P^H also control by NaOH, H_2SO_4, NH_3

(6) Precursor & metabolic regulators :-

- ➢ Addition of such compound support the product formation rather than cell growth.

(a) Precursor :-

- ➢ When they added into medium, they directly incorporated into desired product. Ex.:- Phenyl methyl amine of corn steep liquor can incorporate in penicillin molecules to yield penicillin-G.

(b) Inhibitors:

- ➢ When certain inhibitor is added to media, a specific product may be produced. Ex.: Glycerol production by modifying ethanol production with sodium bisulphate as inhibitor of acetaldehyde.
- ➢ Also help in release of product because increase the permeability of the cell.

(C) Inducer:

- ➢ The majority of industrial interested enzyme is induced by inducer.
  Ex. Starch (Inducer) → Amylase (enzymes)

- ➢ Metabolic intermediate also acts as inducer
  Ex. Fatty Acid → Lipase.

(7) Antifoam Agent:

- ➢ Formation of foam is big problem.
- ➢ Generally produced durign arobic process with highly protein media.
- ➢ Foaming can reduce by following way.
  - (a) Defined medium which give less foam & also modified physical parameter.
  - (b) Adding antifoam agent.
  - (c) Using mechanical foam breaker.

➢ Antifoam agent is surface active agent & reduced surfacetension, ultimately destabilized the foam.

2.2 Ideal Characteristics of Antifoam Agent

➢ Long life activity
➢ Not metabolized by Microbes.
➢ Non toxic for Microbes
➢ Heat sterilizable
➢ Not problematic in product extraction
➢ Should be cheap
➢ Not affect O_2 transfer.

Example: Alcohol, ester, glyceride, fatty acid etc.

Factor involve in fermentor design.
➢ Two problems arise during the designing of fermenter.
  (1) The selection of best type of fermenter for particular reaction.
  (2) The determination of best operating condition.
➢ The selection of design requires good engineering knowledge & judgment with experience.

3. Factor considers designing fermenter.

(1) Selection of organism: - It is important because it will determine……
- The phase in which product is produced.
- The temper & P^H range.
- The degree of aeration.
- Probable affect of contaminants.

(2) Selection of fermenter configuration.
Ex.   --- Batch stirred tank fermenter.
       --- Continuous stirred tank fermenter
       --- Tubular fermenter.

(3) Determination of fermentation dimension.
Volume, diameter, process time, flow rate.

(4) Require of heat transfer surface & mixing devices.

(5) Power & aeration requirement.

(6) Mechanical design:-
       --- Selection of material for construction
           Ex. Steel, iron, wood etc.
       --- Device require for maintenance of aseptic condition.

(7) Facilities for monitoring & control.

(8) Safety factor.

- By knowing above factor, the economic estimation is carried out.
- For this the knowledge & experience of microbiology, biochemistry, thermo dynamics are necessary.
- Following information is also requiring for the design of fermentor.
    (1) The information that associated with the biochemical which take place.
    (2) The rate at which such change is accurse.

(4) Ideal Characteristics of fermenter

(1) The vessel should be operated aseptically for long period of times.

(2) Aeration & Agitation should be provided for best metabolic activities.

(3) Power consumption should as low as possible because it will affect the Overall prize of product.

(4) It should have system to control the P^H & temperature, because as we Know that organism can grow at particular P^H & temperature.

(5) Sampling valve is must be provided to detect the product formation during the process.

(6) Evaporation losses from the fermenter should not be excessive.

(7) The fermenter is designed in such a way that they require minimal use of Labor operation, harvestation & maintenance.

(8) The vessel is designed in such a way that it can be used for more than one types of product formation, but it is not always possible.

(9) The vessel should contain internal smooth surface which is constructed by using welds instead of flange joints.

(10) The vessel should have a similar geometry for both smaller & large vessel.

(11) The cheapest material should be using which gives satisfactory result.

(12) They should provide some additional facilities which gives maximum Product from microbes.

Service provision of fermenter.

- Compressed air
- Chilled water
- Steam
- Motors
- Drainage of influents.
- Electricity etc.

(5) Component of fermenter:

(1) Agitator/impeller:-

Bioreactor contains impeller to ensure a uniform suspension of microbial cell & maintain the homogeneity of nutrient medium.

(2) Aerator / sparger:-

It provides oxygen to submerged microbial culture for their metabolic requirement in the form of air bubbles.

(3) P^H Control System:-

It will maintain the P^H of medium by adding alkali or acid, because microbial metabolic activity requires its specific P^H.

(4) Temperature control system:-

This system will control the temperature so that organism can perform their best metabolic activities.

(5) Antifoam Agent:-

Foam which is produced during process is broken either mechanically or by chemically adding → Antifoam agent.

(6) Baffles:-

It is the metal strip which prevents the vortex formation & improves the aeration efficiency.

➢ Common design of bioreactor is given below.

Fig. 3.1 Ideal Bioreactor

Aeration & Agitation

(1) Introduction:-

- Aeration provides sufficient oxygen to Microbes in submerged culture for their metabolic requirement.
- Agitation is necessary to maintain the homogeneity of nutrient medium.
- The component of fermentor which involved in aeration & agitation.
  - (1) Sparger / Aerator
  - (2) The agitator / impeller
  - (3) Baffles.

(2) Aeration:-

- Majority of Fermentation process are aerobic.

$$C6H12O6 + 6O2 \rightarrow 6CO_2 + 6H_2O.$$

- For complete oxidation of 180 gms of glucose require 192 gms of O_2. Both are soluble but solubility of glucose, is 600 times higher then O_2. Thus it provided during the procedure by means of aeration.
- Factor like O_2 transfer & the factor which influence the O_2 transfer should be studied to determine the O_2 requirement of organism.
- Complete oxidation of carbohydrate substrate equation gives some idea about the oxygen demand of organism for respiration, but not for product & biomass formation.
- The O_2 consumption can be understood by studying the effect of O_2 uptake rate which is shown as QO_2 = milimoles of O_2 consumed / gm dry weigh of cell / hour.
- QO_2 increase with increase in dissolved O_2 (CO_2) up to a certain point, but after critical level it not farther increase.

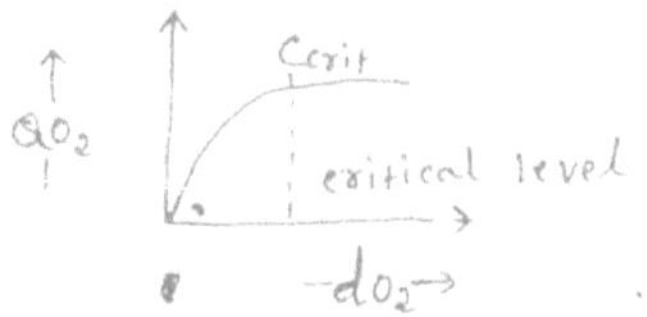

- The maximum biomass production is obtained when we maintained the CO_2 level above the critical level.
- If CO_2 concentration falls below the critical level, the organism is metabolically disturbed.

• O_2 Supply

- O_2 is supplied in the form of sterile air which is cheapest source of O_2
- O_2 transfer from air to microbes occurs in there step.

(1) Transfer of O_2 from air bubble to solution.
(2) Transfer of dissolved O_2 from medium to cell.
(3) Uptake of dissolved O_2 by the microbial cell.

- **The creation system – sparger.**
➢ The devices which are used for the aeration called sparger/aerator that introduced air into liquid of bioreactor.

- **There are three basic types of sparger.**

(1) Porous sparger:-
➢ Made up of sintered glass, ceramics of metal.
➢ Used on lab-scale in non agitated vessels.
➢ Bubbles size is 10-100 times larger than the pore size of aerator block.
➢ Disadvantage: Air come out from this sparger have a low pressure.
   --- Some spore is block due to microbial growth.

(2) Orifice sparger:

➢ Various arrangements of perforated pipes have been tried in different types of fermented vessels with or without impellers.
➢ In small fermented perforated pipes were arranged below the impeller in the form of crosses or rings.
➢ In most design the air holes should drilled on the under surface of the tubes.
➢ The diameter of sparger holes is 6mm.
➢ Orifice sparger has been used in effluent treatment, SCP production & yeast production.

(3) Nozzle sparger
➢ Most recent designed found in lab scale & industrial scale fermenter.
➢ Here open pipes serve as sparger & located below the impeller.

(4) Combined sparger
   It is the combination of agitator & aerator & air supplied through the hollow shaft & emitting from the holes drilled b/w disc.

(5) Agitation
➢ The agitator is used to mix bulk fluid and gas phase, air dispersion, O_2 transfer & heat transfer & maintain the uniform environment within the vessels.
➢ The size, number, speed & power input are specific.
➢ Following are some traditional design of agitator.

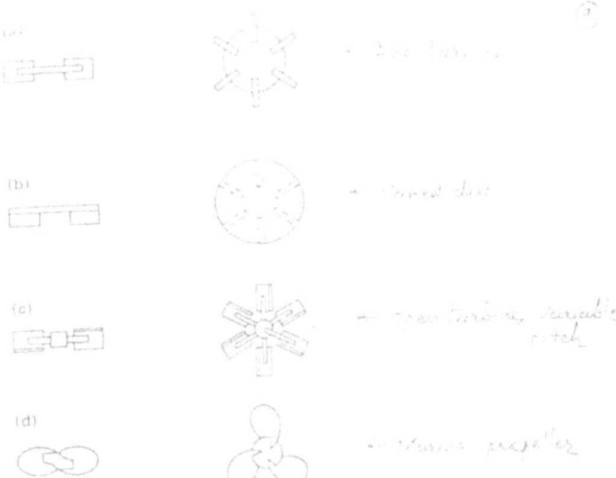

Fig 3.2 Design of agitator

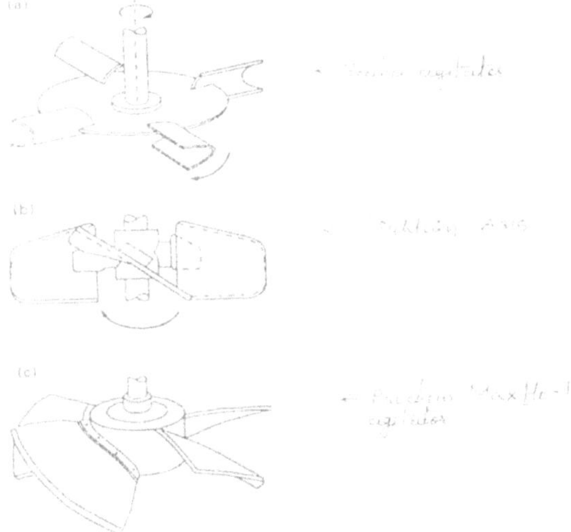

Fig. 3.3 Design of Impeller

- The modern fermenter is generally used in high viscous broth.
- Mainly all of them contain four blades.
- Besides of this interlining agitator made by Ekato of Germany are more complicated in design.

(6) Baffles

> Baffles are the metal strips which are incorporated into the wall of agitated vessels & prevent the vortex formation & improve the aeration efficiency.
> 1/10 of at diameter of vessel.

4. Types of fermentation process

> The condition required for the growth of microbes is differing from the condition that require for the production of particular product.
> Therefore microbes are glow in different way in different vessel, in order to obtained particular product.
> Generally they are glow as batch, fed batch or continuous culture.

4.1 Batch Culture:-

> Bath culture is the simplest method, and commonly used in laboratory for to grown microbes.
> In this method, a desired organism is inoculated into fixed volume of nutrient and allows it to incubate.
> Here there is no addition of new medium or no removal of toxic material during the incubation.
> In batch culture, microbes pass from following growth stages.

(A) Lag Phase:

Initially when we inoculate the organism in the fresh medium, they do not start to multiply within the medium.

> The reason behind this is that microbes required some time for adaptation with new environment called lag phases.
> During lag phase they synthesized the component that required for their growth like enzymes etc.

(B) Log Phase:

After adaptation microbes is enter into log phase.

- In log phase they have all requirements for their growth & sufficient amount of nutrient.
- Therefore they grow very rapidly or exponentially as shown in curve.

(C) Stationary Phase:

- During stationary phase the level of nutrient is decreased to some extent & some toxic material secreted at end of microbial metabolism will not properly support the growth of microbes.
- Therefore number of cell that is generated is equal to the number of cell that is dead and we obtain a static curve.

(D) Death Phase:

- During death phase the amount of toxic material is increased too much around the cell and nutrient too much depleted.
- Here the number of death is higher than number of new generating organism.
- Here only those organism can survive that can fight with such adverse condition & can used to component of death cell as their nutrient.
- Bu, lock et al. proposed different terminology for these growth stages. They used "troposphase" for log phase and "ideophase" for stationary phase. In batch culture this optical growth curve is known as sigmoidal growth curve.

4.2 Continuous Culture techniques:-

- In industrial process it is desirable to maintain bacterial culture to exponential phase.
- This condition is known as steady-state growth.
- Here the volume of culture media & the concentration of cell both arc constant by allowing the fresh medium to enter the culture vessel and at the same rate the spent medium containing cell is removed from the growing culture.
- Under this condition the rate at which new cell are produced in culture medium is exactly balanced by the rate at which cell are being lost through the overflow from the culture vessel.
- There are two method for continues culture.

(a) Chemostate:

- This system depends on the fact that the concentration of an essential nutrient within the culture medium will control the growth rate of cell.
- The con^n of such nutrient is controlled by the dilution rate means the rate at which new fresh medium is added in the vessel. (flow rate)
- Flow rate can be controlled by following equation.

$$D = F/v$$

D = dilution rate.

F = flow rate

V = Culture vessel volume.

- ➤ If the dilution rate is very low then the density of the cell within vessel because they have enough time to utilized the medium and maintains the level of medium at low level. But due to low con^n of substrate the growth rate of cell is very low and vice versa.
- ➤ It means if flow rate is high then the growth rate of cell is high but they have no enough time to utilize the nutrient. The nutrient leave the culture vessel very quickly this is called washout problem.
- ➤ Therefore to prevent the washout problem chemostate is operated at low dilution rate.

(b) Turbidostate:

- ➤ The second types of continuous culture technique is turbidostate in which photosynthetic devices continuously monitoring the cell density within the culture vessel and control the dilution rate to maintained the cell density constant.
- ➤ If cell density becomes too high the dilution rate is increased. If density becomes too low, the dilution rate is decreased.

2.3 Fed Batch techniques

- ➤ Basically it is a batch culture technique in which the nutrient is added at a same rate as they consumed by the growing cell without removing original medium from vessel.
- ➤ This done because, the high con^n of substrate in bath culture will some time inhibit the growth of cell.
- ➤ High density of cell is achieved in this technique as compared to batch culture.
- ➤ Fed-batch culture is an ideal process for the production of intracellular metabolites in maximum amount. Ex. For the production of alkali protease for detergent industry from bacillus species. Here nitrogen source (ammonia) provided by fed-Batch method to induce protease synthesis.